ORANGUTANS

Author Dedication: In memory of Indah, my friend.

Author Acknowledgements:
With thanks to Anne, William, and Carly for their constant enthusiasm and support.
Special thanks are due to Jennifer Wilson and Dr. Serge Wich for generously sharing their expertise and advice.

First published in Great Britain in 2007 by
Colin Baxter Photography Ltd., Grantown-on-Spey, Moray, PH26 3TA, Scotland
Revised edition published 2013

www.colinbaxter.co.uk

Text copyright © 2013 Robert W. Shumaker
Maps © 2013 Colin Baxter Photography based on mapping supplied by Map Resources.
Map data sources: Ancrenaz, M., Lackman-Ancrenaz, I. (2004). Orang-utan Status in Sabah: Distribution and Population Size. Kinabatangan Orang-utan
Conservation Project, Sandakan, Malaysia. Meijaard, E., Dennis, R., Singleton, I. (2004). Borneo Orangutan PHVA Habitat Units: Composite dataset developed
by Meijaard & Dennis (2003) and amended by delegates at the Orangutan PHVA Workshop, Jakarta, January 15-18 2004. Singleton, I., Wich, S., Husson, S.,
Stephens, S., Utami Atmoko, S., Leighton, M., Rosen, N., Traylor-Holzer, K., Lacy, R., Byers, O., eds. (2004). Orangutan Population and Habitat Viability
Assessment: Final Report, IUCN/SSC Conservation and Breeding Specialist Group, Apple Valley, Minnesota.
Revisions to Borneo distribution map on p69: Wich, S.A., Gaveau, D., Abram, N., Ancrenaz, M., Baccini, A., et al., (2012). Understanding the Impacts of
Land-Use Policies on a Threatened Species: Is There a Future for the Bornean Orang-utan? PLoS ONE 7(11): e49142. doi:10.1371/journal.pone.0049142
All rights reserved.

WorldLife Library series

A CIP Catalogue record for this book is available from the British Library.
ISBN 978-1-84107-369-9

Photography copyright © 2013 by:

Front cover © NHPA/Michael Leach
Back cover © D Robert Franz/Franzfoto
Page 1 © David Courtenay/Photolibrary Group
Page 4 © Anup Shah/naturepl.com
Page 6 © Anup Shah/naturepl.com
Page 8 © Anup Shah/naturepl.com
Page 9 © Anup Shah/naturepl.com
Page 10 © David Courtenay/Photolibrary Group
Page 12 © Anup Shah/naturepl.com
Page 13 © Duncan Usher/ardea.com
Page 14 © Anup Shah/naturepl.com
Page 15 © Anup Shah/naturepl.com
Page 16 © Anup Shah/naturepl.com
Page 18 © Anup Shah/naturepl.com
Page 19 © Anup Shah/naturepl.com
Page 21 © Anup Shah/naturepl.com
Page 22 © Anup Shah/naturepl.com
Page 25 © NHPA/Andy Rouse
Page 26 © David Hosking/FLPA
Page 29 © Anup Shah/naturepl.com
Page 30 © David Robinson/Woodfall Wild Images
Page 33 © Anup Shah/naturepl.com
Page 34 © Anup Shah/naturepl.com
Page 37 © Colin Marshall/FLPA
Page 38 © Anup Shah/naturepl.com
Page 41 © Anup Shah/naturepl.com
Page 42 © Anup Shah/naturepl.com
Page 45 © Anup Shah/naturepl.com
Page 46 © NHPA/Alan Williams
Page 48 © NHPA/Mark Bowler
Page 51 © Anup Shah/naturepl.com
Page 52 © Anup Shah/naturepl.com
Page 57 © Anup Shah/naturepl.com
Page 58 © Anup Shah/naturepl.com
Page 61 © Mark Newman/FLPA
Page 63 © Jean Paul Ferrero/ardea.com
Page 64 © Anup Shah/naturepl.com
Page 67 © Anup Shah/naturepl.com
Page 70 © NHPA/Martin Harvey

Printed in China

ORANGUTANS

Robert Shumaker

Colin Baxter Photography, Grantown-on-Spey, Scotland

Contents

Orangutans

I have been captivated by great apes for as long as I can remember. One of my most vivid childhood memories is of watching an orangutan at a zoo. Bars and glass separated us. I stood there in my sneakers and flannel shirt, holding my favorite lunchbox as he sat in front of the obligatory tire on a chain in a stark tiled room. I don't remember seeing his body, but I do recall looking directly at his very large, very round face. What I remember most is his penetrating gaze, fixed firmly on my eyes as I stared at him, and he stared back at me. I was perhaps five or six years old. I wondered what he was thinking.

I suspect that this personal experience was far from unique. Countless other people have reported strong reactions, usually positive but sometimes unnerving, when they look into the eyes of a great ape. Most people aren't prepared for the undeniable sensation that a confident, intelligent, and sentient individual is looking back.

My childhood fascination has endured. For the last thirty years, I've had the privilege to spend my days in the company of great apes, particularly among orangutans. I have learned from them, with them, and about them, and my fascination for these magnificent apes has only increased over time. Although I have a strong and enduring affection for all of the great apes, I freely admit my bias – I've never met an orangutan I didn't like.

All of the great apes have complex mental lives and are behaviorally sophisticated. Orangutans, however, have a special reputation for patience and persistence as they puzzle over the solution to a novel problem. They are notoriously clever when faced with a challenge. Physically, orangutans are

Look into the eyes of an orangutan, and a sentient being looks back.

distinct from the other great apes. Most of their large muscle groups are located in their upper bodies, facilitating styles of locomotion that rely primarily on the strength of their arms rather than their legs. Orangutans are the largest arboreal animal on our planet, spending the majority of their lives high in the forest canopy. It isn't unusual for an orangutan to travel casually through the treetops at heights of one hundred and fifty feet (50 m) above the ground.

Orangutans are the only Asian great ape, and the only ones that don't typically live in large social groups. They're also the only 'redheads'.

While all great apes are endangered, both Sumatran (*Pongo abelii*) and Bornean (*Pongo pygmaeus*) orangutans have one final distinction. Unless the present conditions are substantially changed, they are predicted to be the first great apes to become extinct in recorded history.

This book's job, first and foremost, is to urge orangutan conservation. Long term preservation of sustainable wild populations of orangutans is possible, but only if we change our views of these red apes whose numbers are critically dwindling. In short, we need to understand them, and that means taking a look back, as well as a peek into the future.

The future is uncertain for this adult female and her infant.

Orangutans Discovered – A Historical Perspective

Writings from centuries ago show the undeniable human fascination with apes. One of the earliest of these references was provided by Aristotle, who gave reasonably accurate physical descriptions of apes. He also noted the similarity that existed in the appearances of apes and men. Pliny the Elder (aka G. Plinius Secundus) remarked on the keen mental skills of apes in his *Historie of the World* from 1601. He writes, 'Marveilous craftie and subtle they be to beguile themselves…as they see hunters doe before them, they will imitate them in every point…Mutianus saith, that he hath seene Apes play at chesse and tables; and at first sight they could know nuts made of waxe from others.'

Although we don't know which species are 'apes' in these ancient texts, it is generally accepted that apes were known outside of their native countries as early as 130 – 200 A.D. Of course, within these countries, humans and great apes have co-existed for many thousands of years.

Generally, apes don't appear much in ancient to more recent Western history. All of these writings certainly confused the different kinds of apes, and sometimes perpetuated myths about apes that still influence human perceptions today. Many people still think of apes as violent and aggressive, a stereotype that just isn't true. In Europe, solid evidence of great apes begins to emerge in the 1600's, primarily as a result of Dutch explorers who sailed to Africa and Asia. These men sometimes returned with the remains of apes that had been killed during their travels, and occasionally with a live infant. Not surprisingly, these young apes were an instant sensation, but quickly succumbed to illness or disease. Europeans' only glimpses of live apes were through these infants or juveniles – capturing and transporting a live adult just couldn't be done.

In 1641, a Dutch physician named Tulp recorded the first description of a live ape that had been transported to Europe. Although he called it an orangutan, it is widely accepted that this was a young chimpanzee. In 1658, another Dutchman named de Bondt (also known as Bontius) living in Jakarta, is credited with making the first accurate and reliable recorded description of an

orangutan. After de Bondt, reports of orangutans in scientific and popular literature trickled in slowly. As people learned more about great apes, and orangutans in particular, a few themes emerged. Orangutans were regularly confused with other types of great apes, and natural historians were perplexed by the dramatically different appearance that occurs for orangutans from birth to maturity. They regularly described the intellectual ability of orangutans in positive terms, making early references to tool use and successful problem solving. Given the mental and physical similarities that were noted to exist between humans and orangutans, people naturally began to ask why orangutans couldn't speak.

Writers often applied negative human characteristics to orangutans. Notable examples include reference to orangutans by Linnaeus (1758), Thomas Jefferson in *Notes on Virginia* (1781-1782), and Edgar Allen Poe's gruesome tale of a homicidal orangutan in *Murder in the Rue Morgue* (1841). Poe combines great intelligence with an evil disposition to create his fictional orangutan. As the sailor who owns the orangutan returns home from a night

The first live orangutans to reach Europe were juveniles, such as the
individual pictured here, who had been taken from the wild. These youngsters
were an enormous attraction, but rarely survived for long.

of carousing, Poe writes, '…he found his prisoner occupying his own bed-room, into which he had broken from a closet adjoining, where he had been, as it was thought, securely confined. The beast, razor in hand, and fully lathered, was sitting before a looking-glass, attempting the operation of shaving, in which he had no doubt previously watched his master through a key-hole of the closet. Terrified at the sight of so dangerous a weapon in the possession of an animal so ferocious, as so well able to use it, the man, for some moments, was at a loss what to do. He had been accustomed, however, to quiet the creature, even in its fiercest moods, by the use of a strong wagoner's whip, and to this he now resorted.'

An infant orangutan is totally dependent upon its mother.

These wildly exaggerated tales contributed to the confusion that reigned about orangutans until the last half of the 1800's. Scientists didn't even agree on their taxonomic classification until the 1980's, and our knowledge of orangutans continues to evolve.

Even today, thanks to misleading depictions of orangutans particularly by the television and movie industries, we still don't fully grasp the essence of great apes. Meanwhile, those who dedicate their lives to orangutans and other apes can be as much teachers as scientists, replacing negative stereotypes with understanding and due admiration.

Orangutans: Monkey, Ape, or Primate?

The taxonomy of orangutans is often confusing. Here's a breakdown. All living things on Earth can be divided into functional categories. One of the broadest would be 'plant' or 'animal'. Orangutans, of course, are animals. They also have an internal bony skeleton that includes a backbone (unlike most of the animals on Earth) and are mammals. All mammals have hair, gestate their babies inside of their bodies, and then feed their offspring with milk produced by the mother. Mammals can be subdivided into different Orders, such as bats, or primates (by the way, bats are the Order most closely related to primates). All of the members of a particular Order share similar characteristics. Primates have hands and feet, opposable digits, forward-facing eyes with binocular vision, larger than expected brains relative to their body size, and other physical and behavioral features that should be personally familiar to anyone reading this book. Given our human obsession with categories and sorting, the Order Primate can be further divided into the major groups of prosimians (literally 'pre-monkey'), monkeys, and apes. Common features are present within each of these groups, eventually leading to the narrowest categories of genus, species, and sometimes subspecies. In all, the master taxonomist Colin Groves recognizes about 350 different species of primates.

Apes are classified as Hominoids, which includes the lesser apes (gibbons) and the great apes. All gibbons are lesser apes, a term used to describe their smaller size relative to the great apes. Gibbons share a number of anatomical features with the great apes, but appear to be behaviorally and cognitively less complex. Orangutans, gorillas, bonobos, chimpanzees, and yes, humans are

Similarities between orangutan and human infants are immediately obvious.

classified as great apes, also known as the Family Hominidae.

While all primates share common ancestry, the lineage leading to modern great apes diverged approximately 25 million years ago. Within the Family Hominidae, orangutans were the first to diverge, approximately 14 million years ago, leading to the genus *Pongo*. This was followed by the gorillas, (the aptly named) genus *Gorilla*, approximately 7 million years ago. It's a common misperception that humans were the endpoint of hominid evolution. Not so. The human lineage, genus *Homo*, was the next to diverge from our common ancestry, approximately 6 million years ago. Nearly 3 million years later, the chimpanzees and bonobos diverged, both in the genus *Pan*. While all great apes are closely related as members of the same Family, the closeness of the relationships among great apes is the opposite order of how they emerged. To understand these genetic relationships, consider humans. The closest degree of relatedness, about 98.8%, exists between *Homo* and *Pan*, not *Pan* and *Gorilla*.

Orangutans are one of humans' closing living relatives.

Humans are more closely related to chimpanzees and bonobos than gorillas are. Gorillas share about 98.4% of the same DNA sequences with chimpanzees and bonobos. Humans and orangutans have an overall genetic similarity around 97%. Based on this small difference, orangutans have the highest degree of genetic distinctiveness among the great apes.

Historic and Present Distribution in the Wild

Great apes share significant characteristics, yet each species is unique. Orangutans are distinguished by their historical and present distribution in the wild. Gorillas, bonobos, and chimpanzees are native to Equatorial Africa (also the birthplace of the human lineage). By significant contrast, orangutans have never existed on the continent of Africa, and are native only to Asia.

Historically, orangutans were widely distributed across the Asian continent, from northern India, to southern China, and south to the island

Orangutans may play in water, but are unable to swim.

of Java. The modern distribution of orangutans has been primarily affected by climate, geographical barriers such as waterways, and most likely by human hunting pressures as well. These factors have affected the areas into which these apes have distributed themselves over time. Orangutans can cross slow moving, shallow bodies of water by wading or walking on exposed terrain. However, in the absence of natural bridges such as adjacent tree crowns, they can't swim across wide, deep, or swift moving water. As a result, their historical and present range has been significantly influenced by the presence of bodies of water, such as rivers.

Today, wild orangutans are found only on the two Southeast Asian islands of Borneo and Sumatra. Estimates for the amount of time that these two island populations have been completely separated vary considerably, from about 1.5 million years to 5 million years. The time of their separation has led to differences at the molecular level that are significant enough so that Bornean and Sumatran orangutans are considered separate species. The scientific name for the Sumatran orangutan is *Pongo abelii*, and Bornean orangutans are known as *Pongo pygmaeus*. Anatomical differences within the Bornean population of orangutans, consistent with geographical boundaries made by rivers, are recognized by the presence of three subspecies on that island. The subspecies *Pongo pygmaeus pygmaeus* is found in northwest Kalimantan and Sarawak, *Pongo pygmaeus wurmbii* is located in southwest and central Kalimantan, and *Pongo pygmaeus morio* inhabits northeast Kalimantan into Sabah. In Sumatra, there are no known subspecies of *Pongo abelii*.

Even though there are molecular differences between them, Bornean and Sumatran orangutans easily reproduce with each other in captivity and produce fertile offspring. Also, there are no consistent variations in appearance that allow even an experienced observer to reliably distinguish between Bornean and Sumatran individuals. But, taxonomists can identify differences through detailed examination of anatomical features, such as precise skull measurements. These techniques allow the two species to be reliably classified. Scientists are currently studying and comparing these two island populations to determine if any true differences in behavior and lifestyle can be identified.

All orangutans are behaviorally complex and mentally sophisticated.

Anatomy

What was it like for those first European explorers who ventured on foot into the unspoiled forest home of the orangutan? After sailing halfway around the world, they certainly would have heard the wildly exaggerated tales of monstrous half humans running about in the Asian forest. Did they keep a hand on their weapons, fearing for their lives? Did that grip relax as their eyes adjusted to the soft light of the forest? Still fearful, yet feeling excitement and wonder, too. Towering trees far more than a hundred feet tall loomed overhead, butterflies zagging through the air, sprays of orchid flowers dipping into sight. Perhaps a giant hornbill erupted from a tree hole with noisily flapping wings, disturbed by the intruders near her nesting spot.

Without question, these humans would have looked remarkably out of place. Imagine that very moment when, drawn by this curious sight, a very large adult male orangutan moved to get a better look at of these odd creatures. The orangutan would have seemed immense to these earthbound explorers, almost defying gravity as his bulk traveled from branch to branch. Did they marvel at his body, a beard on his very human-like face? Did their jaws drop at what appeared to be hands on the ends of his legs instead of feet? Before they could react, the orangutan would have vanished back into the forest right before their eyes. If they could believe what they saw, the explorers would now understand the Malay name used for this creature, 'orang hutan', translated literally as 'person of the forest'.

For all of us, the first opportunity to see an adult male orangutan up close and in person is unforgettable. Most people find themselves unprepared for an

This 'person of the forest' has a wise and dignified appearance.

experience that is both humbling and exhilarating. There is an undeniable connection when we look at orangutans and they look back at us. Dramatic physical differences remind us that we are most at home on the ground, while they are much more comfortable skyward.

Orangutans are the largest arboreal animal on Earth, spending most of their time in the trees. They have long lives, up to 60 years old both in the wild and in captivity. They are also the second largest primates. The first are the much more terrestrial gorillas, by a considerable margin. Orangutans, like all primates, share similar physical features. But it is undeniable that they are uniquely designed to spend their lives off of the ground.

Orangutans, like most primates, rely on sight as their primary sense. They have forward-facing eyes with binocular vision. That means they have overlapping axes of vision, which create a three-dimensional image that is interpreted by the brain. Binocular vision is outstanding for distance and depth perception, which comes in very handy if you live and travel dozens of feet above the ground in the forest. One bad decision in the canopy could mean a deadly fall, and excellent vision is required for survival. Being active during the day, orangutans probably see color very much the same way humans and other great apes do. Tropical forests abound in potential foods, but the most preferred foods for orangutans are ripe fruits. There is usually one peak time of the year for fruit availability, although a limited amount of some fruits is almost always available. As a result, a keen eye for the bright color of ripe fruit provides a tremendous advantage when searching the forest for food.

Orangutans process these fruits, and all of the other types of foods that they consume, with powerful jaws and molars that have wide, flat surfaces. These allow them to also consume seeds, bark, assorted types of vegetation, and even some invertebrates. On very rare occasions, orangutans eat meat, but there is no indication that they intentionally hunt, or particularly desire meat as part of their diet.

Orangutans' ability to move about the forest with tremendous grace, agility, and confidence is facilitated by a variety of physical features. The relative proportion of the body, arms and legs is perhaps the most striking of these.

Some orangutans are not particularly tall—adult males are usually about 4.5 feet (1.5 m) tall, and females generally about 3 feet (1 m) tall. Of course, adult height can vary and some males reach nearly 6 feet (2 m) in height. Adult males weigh anywhere from 165 to 290 pounds (75 to 132 kg), and females range from 70 to 150 pounds (32 to 68 kg). Although shorter and

Orangutans are uniquely adapted for life in the trees.

lighter individuals are usually reported in the wild, this is not always the case. Extremes of normal height and corresponding weight are found both in the wild and in captivity. Not surprisingly, taller and heavier sizes are more frequently associated with captivity where highly nutritious foods are consistently available. Some captive orangutans enjoy too much of a good thing, and become far too heavy. Weights from these individuals have not been included here.

The overall design of orangutan bodies is truly remarkable. Orangutans, like

Orangutan hands serve many functions. They are perfectly suited for arboreal locomotion, providing remarkable grip strength high above the forest floor. Elongated fingers are marvelous for plucking ripe fruits, yet also dextrous enough to gently clean the face of an infant.

all apes, lack an external tail. However, evidence for this ancestral primate trait is presented by the coccyx (or tailbone), a vestigial tail found in all great apes, including us. Great ape arms are longer than their legs, but this feature is exaggerated for orangutans. Adults may have an arm span far greater than their height. Taller adult males may measure about 9 feet (2.7 m), or more, from the fingertips on one hand to the fingertips on the other. The same relative dimensions are also true for adult females. Orangutan legs, although certainly strong, are comparatively shorter and weaker than their arms. The larger muscle groups are found in the arms, chest, and back, exactly the opposite of what is found in highly terrestrial species such as *Homo sapiens* whose larger muscle groups naturally occur from the waist down.

Hands and feet are specifically suited to a life spent high in the forest canopy. The fingers, palm, and overall dimension of orangutan hands are extremely long, with a smallish thumb located at the very base of the palm. During locomotion, orangutans use an efficient power grip where the hand acts like a hook, with fingers wrapping around branches and closing with a vise-like grip. An orangutan's fingers are so long that the tips can usually reach all the way round a branch and touch the palm. During travel, the thumb, not involved, simply stays out of the way. Each digit has a fingernail which provides some support for the tips of the fingers during grasping and climbing. A physical detail such as a fingernail might seem minor to us, but primates rely on their nails for numerous functions such as scraping, cleaning, opening, and scratching. Like us, orangutans also have unique fingerprints on each of their digits, allowing any individual to be personally identified from among all other orangutans. The toes are remarkably similar, and allow the feet to be used interchangeably with the hands during locomotion, food collection, play, or virtually any other activity. One important distinction is that the big toe is opposable with the other toes, and is much more useful than the thumb. Hands and feet are large and powerful—think the size of a baseball catcher's mitt. Adults can hang by their arms, without strain, for hours in what appears to be a totally relaxed posture. It's tough to measure just how strong an orangutan is, but we do know that adults can perform one arm 'chin ups' with complete ease. By the way, these can be accomplished while holding on with only one finger.

Incredible strength and coordination, an expansive arm reach, and the equivalent of four equally functional hands with vise like grips, orangutans have skills beyond any Olympic gymnast's wildest dreams. Add tremendous mobility at the shoulder and hip joints, and orangutans are in a category that is unparalleled in the animal world—they are uniquely and astonishingly well suited to a life spent primarily in the trees. To put it plainly, these great apes move like no other species on Earth. Young orangutans in particular have the ability to keep their bodies perfectly upright in mid air by raising each foot next to their head, securely grasping above, freeing the hands to gather food, play, explore, or just relax.

Orangutans travel in a variety of different ways, and are legendary for being able to exploit nearly any surface successfully. This includes wrapping their arms and legs around enormously thick tree trunks and moving up or down, or holding four separate hanging vines and swinging to a new spot with precision. Zoo-living orangutans can navigate smooth concrete walls that defeat professional rock climbers. They have also been known to ascend walls made of glass panels by pressing their hands and feet against the edges in a corner. When moving from one spot to the next, four distinct forms of

*Perfectly at home in the dizzying heights of the forest canopy, orangutans
use skill and good judgment to assess the durability of trees, branches, and vines.
Satisfied with his decision, this unflanged male relaxes in a hammock.*

locomotion are used most often to navigate, and they vary depending on the surfaces that are available. Suspensory locomotion, also called semi-brachiation, allows orangutans to travel along the underside of branches and limbs using only the arms and hands and relying heavily on the rotation of the shoulder joint. During suspensory travel, the body and legs hang below in an upright position. The body is propelled forward as the ape grasps overhead branches one after another. Importantly, during semi-brachiation, one hand is always securely attached to prevent falls. Gibbons, the only true brachiators, move so quickly that they are able to let go completely as they alternate their grasp. That wouldn't be possible for the heavier bodied orangutans. In addition, semi-brachiation only works when an uninterrupted highway of overhead supports is available.

For vertical movement, or when supports are less predictable or more variable, orangutans use their hands and feet in combination. Termed 'quadrumanous' (meaning 'four hands') this style of locomotion uses the hands and feet interchangeably to grasp in any combination, or direction, including upside down. The flexibility of both the shoulder and hip joints is essential for proper quadrumanous locomotion. This style of movement is very safe, and allows for multiple attachment points simultaneously. If a vine held in one hand breaks, the other hand and at least one foot is still elsewhere, greatly reducing the risk of a fall.

On surfaces that are even and predictable such as the ground or large diameter branches, orangutans might choose to walk. This happens in two ways, and individuals are usually consistent in the style they choose. In the first, orangutans use their hands and feet simultaneously. With their very long fingers and toes, this usually means that both the hands and feet are held as fists. The orangutan usually supports its weight on the sides of all four of their fists as they move, never on the knuckles directly (as African apes do). Some individuals might hold their hands and feet open completely, palms and soles flat on the ground, and keeping them at an angle away from their bodies as they move. The second alternative for walking is to use the feet only, and move bipedally as humans do. This can also be done with the feet held in fists or open flat. Orangutans that walk bipedally may do so from habit, or as a way to get a better look around, or to keep the hands free for carrying important items. Without a doubt, walking is the least efficient and least comfortable way for orangutans to move. The adaptations that allow orangutans to be supremely graceful and confident when suspended high in the air are the very same ones that cause them to seem awkward and slow on the ground. Whether in the wild or in captivity, orangutans need to move in the ways that their bodies are adapted to properly operate.

Given these unique physical specializations, it is no wonder that historical depictions of orangutans are misguided. The earliest of these were created by anatomists or physicians who studied preserved specimens; they had never seen a live orangutan. Certainly these educated men noted remarkable similarities between the bodies of humans and those of the orangutans. However, the dimensions of the orangutan bodies were completely novel, and the idea that these apes could spend their lives moving in the forest canopy was unknown, and apparently unimagined, at the time. It was assumed that these apes lived on the ground like humans, and somehow managed to walk about with their oddly proportioned limbs, hands, and feet. As a result, these scientific illustrations presented what seemed to be a logical solution by depicting the orangutans standing erect, and depending on the aid of a cane or walking stick.

Life History

Generally, the lives of primates are dominated by three basic, but very important objectives: finding and consuming enough food to be healthy; producing offspring that will mature and successfully reproduce; and avoiding predation. Even though tigers are a possible threat in the forests of Sumatra, the arboreal lifestyle of orangutans essentially solves this third concern since tigers don't climb trees. As a result, orangutans live within a society that has developed around the priorities of eating and sex – and not necessarily in that order. These two goals exert a strong influence on orangutan natural history and social behavior.

Any form of social living involves a constant balance between the benefits of group living, and the costs of competition. For orangutans, the costs of social living far outweigh potential benefits. Although not properly defined as completely solitary, it is fair to say that orangutans are the least sociable of all the great apes. They do not live in groups, bands, or communities. Adults rarely interact with each other. The only consistent social unit is a mother with her most recent offspring. A variety of factors have contributed to this successful, but most unusual, situation.

Wild orangutans are usually found in habitats that are less than 3300 feet (1000 m) above sea level. Researchers have occasionally found orangutans at higher altitudes, but this is rare as orangutans prefer lowland habitats that are richer in the fleshy fruits that they crave. Lowland habitats vary considerably, ranging from primary dryland forests to swamps; areas that orangutans consider home but that can be inhospitable for humans. Within these areas orangutans are patchily distributed. A common misconception is that orangutans (especially

At birth, infant orangutans may weigh between 3 to 5 pounds (1.4 to 2.3 kg).

males) occupy and defend a discrete territory. Actually, individuals can usually be found within a home range, and multiple orangutans may journey separately through the same area of habitat. In general, travel decisions are heavily influenced by the distribution and abundance of foods, not by territorial boundaries or borders.

The availability of fruit is the single most important ecological factor influencing orangutan natural history. In tropical habitats, different types of fruits such as durian or rambutan, appear throughout the year in a generally predictable pattern. However, depending on local climate conditions, the amount of fruit that is available at any one time can vary considerably. It may be scarce, occur in average amounts, or 'mast' years may provide a super abundant supply. The behavior of orangutans is strongly affected by these fluctuations. When supplies allow, the apes consume fruit almost exclusively. This energy-rich and satisfying food provides a burst of calories, which is especially beneficial after a lean period. Typically, fruits are concentrated in one area, which makes them relatively easy to collect. There is usually very little preparation or processing required before consumption—fruits are natural packages of energy just waiting to be eaten. They provide the orangutans with a tasty meal for relatively little effort. Orangutans require large amounts of food, and fruits keep their weight steady, and their immune systems from being stressed from lack of nourishment.

Even better, eating fruits is always a 'win' situation for the orangutans and sometimes also for the fruits. Some fruit seeds are destroyed during digestion. Others, however, benefit from being swallowed. As the orangutans travel away from a fruiting tree and digest their meal, these eventually end up on the forest floor in a tidy cocoon of organic fertilizer.

A mother and offspring may remain together for nearly 10 years. This long
period of dependence allows youngsters to learn essential survival skills, such as foraging.
Orangutans identify and eat hundreds of different types of foods.

Orangutans are the least social of all the great apes, but should not be considered solitary. The bond between a mother and her infant is the longest and most enduring relationship that exists for these 'people of the forest'.

Not that a diet heavy in fruits is perfect. While very helpful in some ways, fruits provide no long-term benefit in terms of fats or proteins which are essential for a proper diet. Fruit availability is also unreliable. So, orangutans are diverse eaters. In times of plenty, they gorge on preferred fruits, but work by many researchers tells us that orangutans identify and utilize hundreds of different foods, most of them plants or parts of plants. These foods include leaves, seeds, flowers, bark, minerals, insects, eggs, and sometimes small animals. The scientist Anne Russon has even documented a new type of food being utilized by these clever apes. Some individuals living near riverbanks have learned to catch and eat fish.

Orangutans are unquestionably resourceful and inventive; however there are limits on how effectively these skills can address larger issues that exert a significant influence on their natural history. Orangutans are the largest arboreal animal on earth, and rely very heavily on fruits as an essential part of their diet. This can be a problem since fruit in a tropical forest is normally limited, and orangutans need a lot of it. The only possible solution places a limit on the number of orangutans that can take food resources from the same parts of the forest. As a result, orangutans have developed a social system that is unique among all great apes.

Although frequently misidentified as having a 'solitary' mode of existence, it's more accurate to define orangutan society as an extended social system. Take a typical day for an adult female orangutan. As the sun rises, she and her juvenile offspring awake and leave the sleeping nest where they have spent the night. Desirable fruits are difficult to find at the moment, so she decides to travel a considerable distance to a large strangling *Ficus* tree that is a reliable source of figs. Although not a favorite food, these small fruits provide a good source of energy and calories, especially important for a lactating female. As they set off, she hears the call of a familiar adult male in the distance. She notes his location by the distance and direction of the call, but food is the priority today. She proceeds towards the fig tree. Along the way, she stops frequently to rest and eat mouthfuls of tender young leaves that are plentiful and moist. Nearing her destination by midday, she has not seen or heard any other orangutans along the way. She stops and quickly bends and weaves several branches into a modest platform where she and her offspring will rest during the midday heat. As the day cools, they continue traveling until they see the *Ficus*. Moving into the massive canopy of the tree, she sees that a familiar female is present, feeding on figs, and holding a new infant. She has given birth since the last time their paths have crossed. The two adult females move about the tree separately and eat figs until the sun begins to set. With a full belly, our female moves to an adjacent tree where she constructs an elaborate and comfortable night nest where she and her baby will sleep, ready to continue feeding on figs when the sun rises.

Adult female orangutans move through home ranges that, in some cases, have been estimated to be at least 2100 acres (850 hectares) in size. Adult males can range over areas that are up to three times that size, exceeding 6175 acres (2500 hectares). Many individuals may travel through these areas, but as a general rule, adults keep their distance from one another. Orangutans do not defend territories, but they do defend limited resources. The size of a female's home range generally allows her to find sufficient foods for both herself and her dependent offspring. This includes preferred foods, such as fruits, as well as

All adult orangutans construct a new sleeping nest each night.

This young male is beginning to exhibit signs of maturity. Small flanges,
also called cheekpads, are beginning to emerge on the sides of his face. In time,
he will grow to be twice the size of an adult female orangutan.

non-preferred 'fall back' foods when fruits are not available. Typically, this occurs without direct conflict between adult females. Occasionally, some females will chase others away from a fruit tree. It is important to note that food resources are particularly valuable for females when they are pregnant or nursing their offspring. During these times, which may last for several years, females must consume additional calories and so food resources are a high priority for them.

Adult males face a somewhat different set of challenges. They can be twice the size of adult females, and therefore require much larger quantities of food. However, the larger size of their home ranges cannot be explained only by the need for more calories. For males, the additional priority is access to adult females. As a result, the best strategy for males is to occupy a home range that overlaps with as many females as possible, creating more possibilities for copulation and reproduction.

While females primarily distribute themselves based on food resources, males distribute themselves based on females and food. Keeping these factors in mind, the extended social system of orangutans makes perfect sense. Even small groups of orangutans that lived together permanently would have to travel through impossibly large home ranges just to find enough fruit to keep themselves healthy. As a result, adults have a much better chance of making a living if they spend most of their time traveling alone. A single female or a female with dependent offspring can find enough of the right types of food, and so can large bodied adult males with big appetites.

Orangutans also demonstrate some social flexibility. When food is plentiful in the wild, small numbers of females may spend short periods of time together. Adolescent and young adult males may also travel in bands for a few days at a time. In captivity where food is never in short supply, orangutans may be very sociable and even live in groups. With a social system that resolves questions about competition for food, the remaining big issue for orangutans is reproduction – one of the most interesting aspects of orangutan behavior and natural history.

There are two different and sometimes opposing views that are always associated with mating and reproduction, one for males and one for females. In order to understand orangutans, it's important to consider both.

Male orangutans spend the overwhelming majority of their time alone, traveling through a home range foraging for food, and hopefully finding or attracting sexually receptive females. When copulations occur between a pair, the male and female may spend from a few days to a few weeks together before they separate and resume their independent lives.

Adult females also travel through a home range foraging for food, but have more opportunities to mate, either by being approached and pursued by males, or choosing a male that they prefer.

The critical difference in male and female reproductive strategies is that males are in constant and direct competition with each other for mates while female competition is far less direct. To succeed, males can either out compete each other with size and strength, or by being inconspicuous and pursuing females. Orangutan males utilize both strategies, but not by behavior alone.

Unique among all mammals, male orangutans actually appear in two distinct forms. One has all of the physical features that blatantly advertise 'maleness', while the other retains a more adolescent, androgynous appearance. Both types of males are fertile, and both actively pursue sexual opportunities with females.

Females have clear preferences for some males over others, and this mate choice exerts a very important selection pressure on males. Given that females

don't often associate with adult males, they have little behavioral information to judge their choice of a sexual partner. Fully developed males, therefore, rely heavily on their looks to impress females and intimidate rivals. To compete with each other, adult males with full secondary sexual characteristics are usually at least twice the size of adult females. A larger body provides an obvious advantage when trying to prevent sexual rivals from approaching desirable females, or during direct aggression with another male. These differences in appearance between males and females are not only about male-male interaction, females prefer larger males with better developed physical features.

The size to which adult male orangutans can grow is held in check by two factors. The first is physical compatibility with females, and the second limitation is the amount of food that must be found in order to sustain a huge body.

Adult males that have fully expressed secondary sexual characteristics are truly spectacular. In addition to impressive body size, they have other physical features that exaggerate their large appearance. The red hair covering their bodies can be quite long, cascading down their bodies and doubling their silhouette. Large cheekpads, or flanges, adorn either side of the head in a half-moon shape just in front of their ears. Cheekpads vary somewhat in form from male to male, but always lend the appearance of a much larger, wider head. Females may judge a male's overall health by these cheekpads since they decrease in size along with weight loss and old age.

Male orangutans also share an interesting distinction with human males, both sport fully developed beards when reaching sexual maturity. Males also have a large, pendulous throat sac that acts as a resonance chamber for the

In captivity, body size may become exaggerated due to obesity.

booming long-call vocalizations that only these males produce. These swaggering announcements can be heard from great distances, advertising a male's location. Rival males may heed the message and keep a safe distance, or approach to engage in aggressive conflict. Listening females can follow the calls to find males they prefer, and avoid males they don't—strongly suggesting that males can be individually recognized by their voice and call.

The whole package of macho characteristics emerges simultaneously, and development of these begins anywhere from the mid teens to the mid thirties. Given the tremendous competition that exists between these fully developed males, there are no known cases where they have ever been socially compatible, either in the wild or in captivity. Interactions between these males involve either avoidance or aggression—there appears to be no middle ground.

A somewhat different social scenario exists for 'unflanged males', the fertile males who don't have these 'macho' sexual traits. These orangutans are sexually mature, yet lacking in all of the physical and behavioral features that characterize fully flanged males. This male form may include individuals across a range of ages, from early teens to nearly thirty years old. They also lack the same social restrictions that exist for flanged males, sometimes associating with each other in small cohorts. Although these individuals are dominated by flanged males, a degree of tolerance can also

Injuries are common during male fights.

exist. Unflanged males do not normally make long-calls, and may share the same general area as a flanged male while remaining relatively inconspicuous. This strategy also allows them to be near females, who may be available for copulation. Females prefer fully flanged males, and rarely solicit the sexual attention of unflanged males who must rely on a different reproductive strategy. These smaller and lighter males actively pursue females, something that is more difficult for the much larger and heavier flanged males. Females may be followed and harassed until they allow copulation, or forced copulations may also occur. This does sometimes happen with flanged males, but is more common with unflanged males.

Bimaturism, the term that describes these two different forms of male appearance and behavior, occurs both in the wild and in captivity, although the mechanism that produces this effect is not well understood. Scientists think that the presence of a male with full secondary sexual characteristics can suppress the development of other males, causing them to remain unflanged. It may be that this phenomenon is associated with pheromones, hearing long calls nearby, or other unidentified causes. If a flanged male permanently leaves an area, an unflanged male can rapidly develop cheekpads, long hair, and begin to make long calls. This dramatic transition occurs over the course of about a year. Both flanged and unflanged males regularly reproduce in the wild, so both reproductive strategies are effective for males.

The path to female sexual maturity is less complicated. As with males, fertility can occur before the tenth birthday, but the physical appearance of adulthood may not develop until the late teens or early twenties. All females have similar physical characteristics upon reaching maturity; there is no such thing as bimaturism for them. A female's first birth usually occurs around the age of 15 or 16 years old, and her next baby will not usually be born until 6 to 9 years later. This interbirth interval is the slowest of all the primates, including humans. In fact, it is the slowest rate of reproduction for all land mammals. Over the course of her lifetime, a female will be very fortunate to have 4 or 5 surviving offspring. Menopause has not been documented in orangutans.

Infants are completely dependent on their mothers when they are born, and have a very slow rate of development. This dependency lasts for years and young orangutans simply will not survive without the devoted care of their mother. In the wild, adult males provide no care for either infants or juveniles. In fact, adult females typically avoid interactions with males when their offspring are very young. In captive settings, adult males show little interest in infants, but may be very tolerant and particularly playful with juveniles and young adolescents. Although it has been documented in gorillas and chimpanzees, there is no evidence for infanticide in orangutans, whether in captivity or in the wild.

The strong interdependence between mother and offspring testifies to at least two important features of orangutans. The first is the intense physical and emotional bond between a female orangutan and her baby. This bond certainly also exists for the other great apes, but in social settings that include interactions with many other group members such as aunts, cousins, grandmothers, and other youngsters. This extended family group is absent for the orangutan mother-infant pair. The second is the tremendous amount of information that must be absorbed by a young orangutan from its mother before being able to survive independently. This learning process demonstrates the complexity of life in the wild for orangutans, as well as the impressive mental abilities that are required to decipher, retain, and flexibly utilize information collected over the course of many years.

Behavior and Cognition

Orangutans, like all great apes, have brains that are larger than expected for mammals of their size—generally true for species that have exceptionally long periods of youthful dependence. For orangutans, it may be almost a decade before youngsters are fully independent. To become self-sufficient, orangutans use their mental abilities to learn as they grow. With Mom as a model, young orangutans learn skills such as foraging, traveling, and social skills such as who to approach and who to avoid. For example, as a mother orangutan collects and eats figs, her youngster sees this food source, touches it, or even pulls them directly from a stem. Mom may share bits of chewed fig, passing it directly from her lips into the mouth of her curious baby. In time, her offspring will begin to pick up familiar foods and eat them directly. Gradually, these foods become part of an ever growing and exceedingly long list of edible items that will be utilized over a lifetime. Orangutans have also demonstrated more complex forms of social learning, such as imitation of a new behavior. Imitation is a particularly efficient form of learning that requires mental sophistication. If an individual can imitate another's behavior, it implies that the whole behavior, from beginning to solution, has been understood and can be duplicated. Therefore, the potentially long and frustrating process of trial and error learning can be avoided.

Orangutans in the wild and in captivity have demonstrated keen ability to imitate. For example, after watching their caretakers, zoo-living orangutans have learned to sweep any trash off of the floor, mix soap and water in a bucket to make lather, dip in a brush, scrub the floor, dump out their bucket, and then rinse the floor with clean water. Orangutans have shown remarkable diversity

A youngster learns by copying the technique for drinking raindrops.

in their imitative actions, such as hanging up a hammock to rest in, sharpening an axe blade, or attempting to make a fire. These specific examples have little to do with the life of a wild orangutan but they illustrate how successfully new behaviors might flow from one wild orangutan to another.

But, because they spend so much time high in the trees, and because their habitat can sometimes be forbidding, it can be tough for humans to test orangutans' cognitive abilities in the wild. Much of what we know about cognitive abilities has been studied in captivity. When presented with a mirror, orangutans routinely demonstrate that they understand their reflection. No other primates except great apes have convincingly shown this ability. This sense of self has also been explored with other studies. Orangutans have convincingly shown that they have a 'theory of mind', meaning that they can understand the world from another individual's perspective.

Early explorers and anatomists wondered if orangutans could talk, and a Dayak legend holds that orangutans have always known how to talk, but they refuse to do it for fear of being put to work. Scientists are still exploring questions about language. Orangutans can learn to use American Sign Language confidently and creatively (a male orangutan named Chantek proved that at the University of Tennessee at Chattanooga many years ago). More recently, other orangutans have learned to understand abstract written symbols on a computer, and they can use those to communicate their thoughts, desires, and perceptions about the world.

Of all the abilities that have been studied with great apes, orangutans have gained a most impressive reputation for tool use and tool making. From the earliest studies, orangutans appear to have a natural inclination for solving problems with tools. This may be related to the fact that orangutans are contemplative, highly attentive to details, and extraordinarily patient. They appear to have an endless fascination for solving problems that are presented to them.

Orangutans have used tools for countless tasks, such as sponging, hammering, reaching, hitting, prying, or digging. They have made tools that serve as cups, plates, hammocks, rain umbrellas, brooms, or ponchos. Orangutans have been observed weaving fibers into a rope that they then use, and have been rumored to pick locks with pieces of wire that they have found and hidden until no one was watching. These lists are far from complete and only hint at the range of tool using and tool making abilities that have been documented for orangutans. They show tremendous diversity and creativity in their modes of tool use.

Field researchers who have studied wild orangutans at seven different sites across Borneo and Sumatra have compared their information, and found many examples of tool use, as well as other types of behaviors, that vary from one site to the next. The first tool using behavior that was noticed as being different among sites involved the *Neesia* fruit. These are oval shaped fruits ranging in size from a grapefruit to a large pineapple that have very sharp spines on the outside. The inside has oily seeds that the orangutans find delicious, but they are protected by stinging hairs that are like fiberglass insulation. As *Neesia* ripen, thin openings appear on the sides. The challenge for the apes is finding a way to remove the seeds from the spiky fruit without getting stinging hairs on their fingers or inside of their mouth. Only some of the populations of orangutans that have access to this fruit have learned how to eat the seeds. At these sites, they insert slender sticks into the openings, and knock the seeds out so that they fall into their mouths. With this technique, they can avoid the stinging hairs while eating the nutritious seeds.

However, other unique forms of tool use were found when these seven sites were compared, such as using leaf napkins to wipe sticky food off of the face. In some cases, orangutans at separate sites solved a similar problem with the same tools, indicating that this innovation had been invented at each of these locations and then spread within the local population—such as apes using leaves like gloves when crossing a spiny branch, or using leaves as a cushion to sit on a spiny branch. One rare behavior that was found at two sites involved orangutans holding a bundle of leaves, like a doll, while they slept in their nest at night.

The researchers documented 43 behaviors that differed among the sites. Importantly, as the geographic distance increased between sites, so did the variation in behaviors. Also, at sites where orangutans came into contact with each other more often, there were more distinct behaviors with greater opportunities for social learning. All of these factors support the conclusion that culture is the best explanation for the variation in behaviors identified among these distinct populations of orangutans. Currently, extensive cultural variation has only been documented to occur in humans, chimpanzees, and orangutans.

Research from both captivity and the wild reveals that orangutans have a sense of self, are experts at tool use and tool manufacture, can learn to use language, have the ability to understand things from another individual's perspective, learn from each other, and demonstrate cultural variation among different populations. We have just begun to understand the minds of orangutans, and some of the most impressive results have only recently been discovered. As research continues into the cognitive abilities of these great apes, we should be prepared to be amazed at what will be revealed.

Orangutans are masterful tool makers. A large leaf provides relief from the sun.

Conservation

As a young child, my introduction to orangutans sparked a lifelong passion. As a grown man with greater knowledge and understanding, my view of orangutans is basically unchanged. I view them with wonder and amazement. In so many ways, we are like them and they are like us. I have had the unusual privilege to form life-long friendships with orangutans, sharing with them the incredible joys and deepest sorrows that inevitably come with living a full life. For me they are far more than a topic to study, or a subject to teach. Orangutans have likes and dislikes, distinct personalities, and varying intellectual abilities. Stated most simply, they are individuals who deserve far better treatment than our species is currently providing for them, in captivity and particularly in the wild. It is unthinkable to me that my young son and daughter could become adults in a world without wild orangutans; they may disappear from the forests of Borneo and Sumatra well within my own lifetime.

Perhaps the most frequently noted generalization about orangutans is that, compared with the other great apes, their pace of life is slow. So far, almost everything we know about orangutan natural history and behavior supports this idea, including the rate at which females reproduce. There is a single exception to this rule, and it is the rate at which orangutans are disappearing from the wild and rapidly heading towards extinction.

Orangutans are among the most endangered of all the primates. The numbers of wild orangutans have been steadily decreasing for decades, from hundreds of thousands in the early 1900s to a total of about sixty thousand today. The most current numbers estimate approximately 6,600 orangutans living in Sumatra, and about 50,000 living in Borneo. These numbers are absolutely tied to remaining forest, which is the critical factor for the survival of orangutans in the wild. Most importantly, forested areas have been disrupted throughout Sumatra and Borneo, and the remaining orangutans survive in increasingly isolated patches of habitat. Given that orangutans have the slowest rate of reproduction for all primates; populations of less than 250 individuals are not sustainable into the future.

Sumatra provides the best illustration of the current situation. There are 10 separate populations of orangutans left in the northern tip of the island, and only six of those have at least 250 individuals. Those six populations remain threatened with continued deforestation of their habitat. Two examples demonstrate the magnitude of the crisis. In the first, the Tripa peat swamp forests on the west coast of Aceh have been seriously degraded since 2007, and the future of that population remains uncertain at best. Extinction of the Tripa orangutans remains a real possibility. The second case involves huge areas of lowland forest in the Aceh Province where most Sumatran orangutans occur. This critical habitat is in serious jeopardy due to efforts aimed at new concessions for logging, conversion to palm oil plantations, and mining. If these threats become reality, the effects for the Sumatran orangutans will be devastating. Scientists predict that unless current conditions are reversed, 50 percent of Sumatra's orangutans will be gone in ten years, and the whole population could be lost within one generation.

There are more Bornean orangutans, but as in Sumatra, their habitat is disappearing at an alarming rate of at least 10 percent per year. If current trends continue for Borneo, scientists predict that these orangutans could also be lost within one generation. If these predictions do come true, orangutans will be the first great ape to become extinct in recorded history.

Threats to Survival

The activities that threaten the survival of orangutan are easy to identify, but very difficult to stop. The major threats are logging, agriculture, and hunting. Tropical

hardwood trees are highly prized for their beautiful and durable wood, yet these trees take many years to grow and are difficult to produce as a crop. The revenue that can be earned from just one of these trees provides a strong motivation to harvest them from the forest, even if it's done illegally. Faster growing, more common trees can also be an important source of cash when sold for building materials. Frequently, the wood from these trees is exported to be sold in other countries in both the East and the West.

If not terribly damaged, a forest does have the ability to recover if left untouched. Once an area has been logged, however, it is usually converted for other uses such as agriculture. Commercial agriculture can be a positive influence for a local economy by providing jobs and a steady income for workers. Unfortunately, tropical forest is being replaced with enormous plantations, such as those used to grow palm oil. Instead of cultivating abandoned areas or implementing better farming practices on existing agricultural land, profits from logging a pristine forest are used to convert the area for agricultural purposes. This strategy is far more profitable, but also extremely damaging to orangutan conservation.

As roads are built for logging or agriculture, the number of people in or near the forest steadily increases. Simultaneously, apes are losing habitat and with that, their food becomes harder to find. Conflicts between humans and orangutans arise, which apes inevitably lose. People may feel threatened by the orangutans, and react by injuring or killing them.

Also, poachers may take advantage of increased forest access and begin to actively hunt for orangutans, especially females with babies. Unlike the situation currently facing the African apes, the commercial use of orangutans as meat is rare. However, babies can be sold for a profit, and the illegal trade in private ownership of apes is flourishing in Indonesia, and other parts of Southeast Asia. The only way to capture a very young orangutan is to kill the mother, and take the clinging baby from her body. One can only imagine the trauma that this inflicts on the terrified, dependent baby.

In addition to the deaths of the mothers, it is estimated that three or four babies die for every one that survives to be sold into the illegal pet trade, or used as some form of attraction or entertainment. Living conditions for these young orangutans vary tremendously. Some are treated as surrogate children, while others are kept in horrific circumstances. Inevitably, as they grow, it becomes impossible to keep them in a home. Many are abandoned, die, or wind up living alone in a barren cage. If they are lucky, they will be confiscated by authorities and sent to a proper rehabilitation center.

Hope for the Future?

Although the current situation for orangutans is bleak, there is hope. In some cases, logging and agricultural companies coordinate their activities with orangutan advocates to make the best of a difficult situation. As areas of forest are planned for logging, teams from rehabilitation centers are notified so that they can attempt to relocate orangutans into other areas. The same may be true for orangutans that are displaced from their forest home and find themselves surrounded by commercial agriculture. This is not a long term solution to the larger problem, but it does provide a chance for individual orangutans who are

All orangutans are in danger of extinction – especially those remaining in Sumatra.
The time to act is now. Unless current trends are quickly reversed, this may
literally be the last generation of orangutans to grace the forests of Asia.

facing a life-threatening challenge.

Dedicated people continue their heroic efforts to confiscate and rehabilitate the large numbers of babies that are still bought and sold illegally for private ownership. Hundreds of these orphans are now living in rehabilitation centers on Borneo and Sumatra. There is a steady flow of new orphans into these centers as the forest continues to disappear. Options for releasing these young orangutans back into suitable habitat are limited now, and becoming increasingly rare. The most urgent need is improved protection of areas were wild orangutans continue to survive, ensuring their long term conservation.

Looking towards the future, attempts are also being made to reforest land that could eventually become proper habitat, especially where it would reconnect patches of forest that are now isolated. Moreover, conservationists are increasingly active in finding ways to provide economic alternatives for local people who might be tempted to engage in illegal logging or poaching of orangutans. Effective conservation must honorably and respectfully include the interests of local people so that they may pursue a dignified quality of life. Critically, government officials must enforce existing laws against poaching and illegal logging in order to stop the current spiral of destruction.

How to Help

Many consider the current situation for orangutans to be one of the worst conservation failures of our time. That may be true, but there is still time to reverse the trend and stabilize current populations. The first step is being informed—and reading this book and others like it is a good start. Other reliable sources for current information are provided at the end of this chapter.

The most up to date and important information is collected by scientists living and working in range countries. In addition, the mere presence of researchers that study orangutans is a very positive influence on conservation. A research camp can effectively deter illegal logging and poaching. These projects also employ local people as research assistants, trackers, camp workers, and guards against poachers. The employment and educational opportunities associated with a research site create enthusiasm for conservation among local people. Researchers that devote themselves to the study of wild orangutans deserve strong and reliable financial support, not only from businesses and educational institutions, but from individuals as well. Members of the public can donate directly, attend lectures presented by field workers, buy their books, or potentially volunteer their time.

Making conservation-minded purchasing decisions also helps. We live in a global economy, and how people spend their money matters. It is perfectly appropriate to ask merchants if they can guarantee that their products were produced or harvested in a legal and sustainable way. For example, 'tropical hardwood' for furniture may have been harvested from an Indonesian forest, directly affecting orangutans. The responsibility for orangutan conservation includes everyone, not just the people living in countries with wild orangutans.

Getting involved in an orangutan conservation group is another way to act for change. Get in contact with organizations that promote protection for orangutans (see the 'Useful Contacts' list), and ask for more details. While financial support is always appreciated, personal interest is highly valued. The best way to contribute may be through volunteering, or simply by spreading accurate information.

It is no exaggeration to say that orangutans are facing a conservation crisis, and they desperately need our assistance. By conserving the forest where orangutans live, thousand of other species are also saved from extinction. The actions of a single individual can have a powerful and long-lasting effect, and that can begin with you.

Historical Distribution of Orangutans

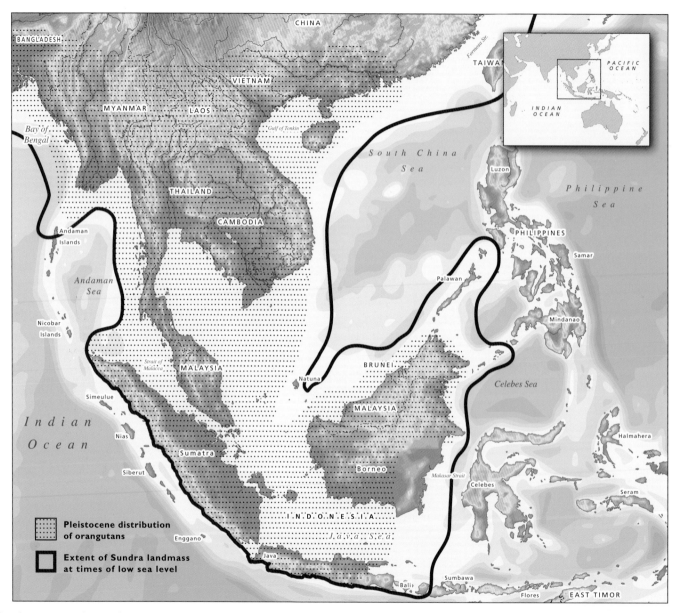

Fossil remains indicate that orangutans existed throughout Asia in prehistoric times. Their range extended into northern India, southern China, and Java. At the end of the Pleistocene, changes in land masses, forests, and waterways, influenced their modern distribution into Borneo and Sumatra. Since orangutans don't swim, this required travel across naturally occurring bridges. *(map courtesy of Dr. Herman D. Rijksen)*

Present Distribution of Orangutans

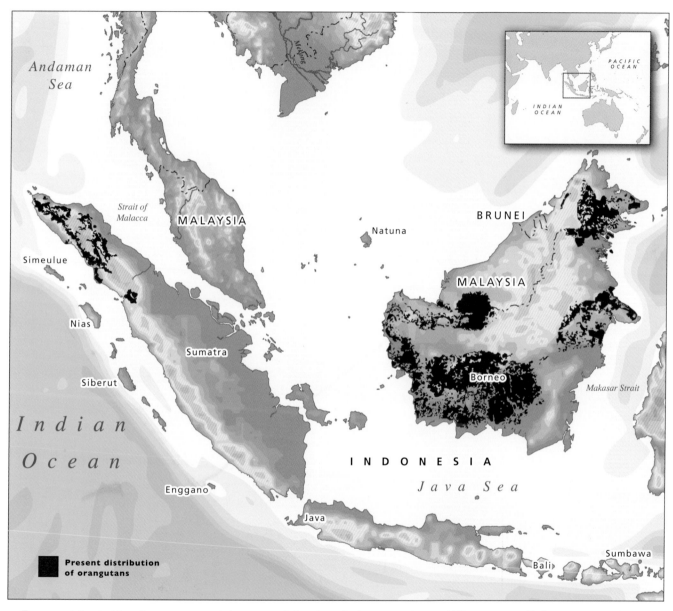

The remaining habitat for orangutans is but a small fraction of what existed even 100 years ago. Today, most populations of orangutans live in 'islands' of forest habitat surrounded by humans – limiting or preventing normal ranging patterns. As a result of human activities, critical areas of forest continue to be lost. Habitat preservation is essential for orangutan conservation.

Orangutan Facts

Species:	Sumatran Orangutan *Pongo abelii*	**Species:**	Bornean Orangutan *Pongo pygmaeus*

Subspecies: *P. p. wurmbii*
P. p. pygmaeus
P. p. morio

Weights: Adult males in the wild, 190 pounds (86 kg)

Adult females in the wild, 88 pounds (40 kg)

Gestation: 250 – 270 days (8.3 – 9 months)

Birth Weight: 3 to 5 pounds (1.4 to 2.3 kg)

Birth Interval / Age of Weaning: 6 to 9 years

Female Age at First Birth: 15 years

Longevity: 60 years

Social System: Variable, but not truly solitary. Adult males and females may spend short periods of time together.

Range Size: Variable, influenced by quality of habitat.
Estimated minimum for adult females: 2100 acres (850 ha)
Estimated minimum for adult males: 6175 acres (2500 ha)

Cultural Variation: Documented for both Bornean and Sumatran orangutans.

Useful Contacts

The following organizations can provide additional information about orangutans and their conservation status.

Indianapolis Zoo
1200 W. Washington St.
Indianapolis, IN 46222-0309, USA
Tel: +1-317-630-2001
www.indyzoo.com

Orangutan Conservancy
P.O. Box 513
5001 Wilshire Blvd. #112,
Los Angeles, CA 90036, USA
www.orangutan.com

Sumatran Orangutan Conservation Programme
Jl. K.H. Wahid Hayim No 51/74, Medan Baru,
Medan 20154, Sumatera Utara, Indonesia
Tel: +62-61-451-4360
www.sumatranorangutan.org

Borneo Orangutan Survival Foundation
Jalan Pepaya No. 40, RT003/05,
Jagakarsa, Jakarta 12620
Tel: +62-21-70721926, 7874479
www.orangutan.or.id

Borneo Orangutan Society, Canada
74 Boultbee Ave., Toronto ON M4J1B1, Canada
Tel: +1 416 462 1039   www.orangutan.ca

Orangutan Foundation U.K.
7 Kent Terrace, London,
NW1 4RP, UK
Tel: +44-207-724-2912
www.orangutan.org.uk

Leuser International Foundation
Jl. Bioteknologi No. 2 Komplex USU
Medan, Sumatera Utara, Indonesia
Tel: +62-61-821-6800
www.leuserfoundation.org

Index

Recommended Reading

World Atlas of Great Apes and their Conservation (University of California Press, Berkeley, 2005) edited by Julian Caldecott and Lera Miles. This volume is a comprehensive description of what is currently known about all of the great apes and their conservation status.

Orangutans; Wizards of the Rain Forest (Firefly Books, New York, 2000) by Anne E. Russon. This volume describes the natural history of the orangutan, and specifically focuses on the learning processes associated with the rehabilitation and reintroduction processes.

Primates in Question: The Smithsonian Answer Book (Smithsonian Institution Press, Washington, D. C., 2003) by Robert W. Shumaker and Benjamin B. Beck. This book provides an approachable and interesting introduction for anyone interested in learning about primates.

Among Orangutans; Red Apes and the Rise of Human Culture (The Belknap Press of Harvard University Press, Cambridge, 2004) by Carel van Schaik. The author provides a thorough overview of the complex behaviors of wild orangutans, and the emergence of culture.

Biographical Note

Robert W. Shumaker is an evolutionary biologist who has worked with orangutans for over 30 years. His research interests focus on comparative cognition, with an emphasis on orangutans. He has long been a strong advocate for animal welfare in captivity and conservation in the wild. Dr. Shumaker is the Vice President for Conservation and Life Sciences at the Indianapolis Zoo. He is an adjunct faculty member at IU in Bloomington and an external research associate at the Krasnow Institute of George Mason University. Other books include "Primates in Question: the Smithsonian Answer Book", and "Animal Tool Behavior: The Use and Manufacture of Tools by Animals".